Basic
WOOD BURNING

with Sue Waters

Text written with and photography by Joanne Tobey

Schiffer Publishing Ltd

77 Lower Valley Road, Atglen, PA 19310

Contents

Published by Schiffer Publishing, Ltd.
77 Lower Valley Road
Atglen, PA 19310
Please write for a free catalog.
This book may be purchased from the publisher.
Please include $2.95 postage.
Try your bookstore first.

We are interested in hearing from authors
with book ideas on related subjects.

Printed in China.
ISBN: 0-88740-568-1

Introduction

The purpose of this book is to help you get started—with as few problems as possible—on the enjoyable and interesting art form of wood burning. Within these pages, you'll find everything you need to know to create your own wood burnings. Included is information on choosing a wood burning tool, the types of wood and other supplies you'll need, wood preparation, tracing and transferring a pattern, and lettering. I've included the patterns and step-by-step instructions for producing six beautiful wood burnings—a starting point from which you can create your own special designs.

Although I've written the book primarily for the beginner, I've included a lot of useful information for the intermediate artist; special tips for making the most of your equipment and for getting the best results when you wood burn.

At whatever level you pursue your wood burning, I hope you find the art as much fun as I do.

Supplies

Wood burning tool (single unit or solid state)
Wood (bass-wood plaques or plates, or bark-on planks, available at craft stores or from cut lumber)
Sandpaper or sanding disc
Tracing paper
Graphite paper
Tape
Soft-lead pencil
Metal-edged ruler (optional)
Varnish (non-yellowing, brush-on. I prefer Satin Finish Right Step, available at craft stores.)

Tools

There are two different types of wood burning tools: the single hand-held unit that most of us are familiar with (it looks like a large plastic pen with an electrical cord at one end and a brass burning tip at the other), and the solid-state machine with temperature control and changeable hand and tip pieces (the "tip" is what you actually use for "drawing"). When I first began to wood burn, I used a single-unit tool; I advanced to the solid state when I wanted to work more quickly, and you'll see it featured in the pictures in this book. Either tool is fine for your first efforts; the single unit is the better choice for children, who will find it safer and easier to use.

The single-unit tool generally has one versatile tip (there are some models with interchangeable tips) and burns at one temperature. The depth of the burn is determined by the speed at which you work. For a dark burn, work slowly; for a lighter burn, move the tip more quickly.

It takes five to ten minutes for a single-unit tool to completely heat. It also takes that long to completely cool down. Placing the tool on a ceramic tile during heat up, cool down, and when on but not in use is a good safety practice. One wants to burn a picture, not the house.

Solid state units have a temperature control that allows you to use very low or very high heat—and anything in between. I usually set the temperature somewhere in the middle, varying it occasionally, depending on the technique that I'm using and whether or not I want a dark or a light burn. I always use a low heat setting when I'm doing small, intricate work, so that I have better control over what's happening on the board.

There are many tips that can be used with a solid-state unit. For the purposes of this book, I'm using two of the most common tips: the shading tip and the writing tip. The writing tip is a thick, curved wire, with a surface area similar to that of a pen; it burns a relatively fine line. The shading tip has a "spoon" on the end of it: a broad surface that allows you to burn a fairly wide area.

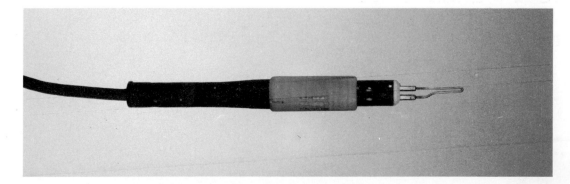

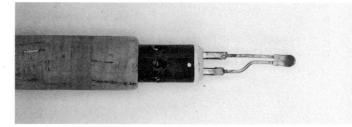

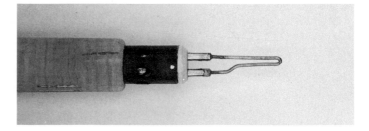

Wood

There are two types of wood that I favor for flat wood burning: birch plywood and bass wood. Both are light-colored woods, with smooth grain that's easy to burn over. (If you try to burn on a rough-grained wood, such as oak, you'll find that your burning tip catches on the grain, producing a very uneven line. Pine has a pitch that makes burning difficult, and some woods are so dark that the burn does not show well.) Birch plywood is readily available at most lumber yards; it's easy to cut the board into picture sizes, using a hand saw or circular saw. Bass wood usually comes in planks up to fourteen inches wide and three-quarters of an inch thick. This wood is great for making your own plaques. You can cut the board to size and rout the edges to give them detailing. Most turned plates (available at craft stores) are cut from bass wood.

Preparation of the Wood

Regardless of the type of wood you're using, it's very important to sand it before you sketch out the piece or begin burning. The smoothness of the wood allows the burning tip to move easily over the surface. Rough wood tends to snag the tip, leaving a dark burn where you might not want one. Working with the grain of the wood, sand lightly until you have a silky smooth surface. If your wood is very rough, start with heavy grit sand paper, working your way up to the lightest grit.

Lettering

The lettering on each plaque in this book was done with stencils that can be bought at art- or business-supply stores. For your convenience, I've included the letters used so that you can personalize your own plaques.

You may notice that the lettering on the patterns is different from that on the finished plaques. This is simply due to a quirk of mine. I prefer a more closed letter than a stencil allows, so I fill in the gaps free hand. Feel free to do the same on your plaques.

The dots below the stencil letters are to help you with spacing. When you trace off a series of letters for a name or word, trace the dots as well. Match the left-hand dot of the letter you are tracing with the right-hand dot of the previous letter, and the letters will be evenly spaced.

TROUBLESHOOTING TIPS:

Before using the single-unit burning tip for the first time lightly sand its edges and tip to remove the sharp edges Doing so allows the tip to glide over the wood without gouging it. This is not necessary for solid state-tips.

Use a ruler to make straight lines on your tracing paper before you start tracing the letters. If you plan to use several lines of letters, rule all of the lines before you start your tracing. This will keep your lettering neat and in line.

Don't work in the breeze of a fan or air conditioner. The moving air will cool your burning tip, causing an unevenness in your burning.

If your burning tip starts to drag, you've got carbon build-up. Make your work go more smoothly by sanding the tip lightly with a fine-grit paper or disc.

Don't wear hand lotion while you're burning, or the oil in the lotion will stain your board.

If you have long hair, tie it back while you're working, or you may burn the tips!

A wood burning is like a fine water color. Keep it out of the sun or the wood will darken, causing you to lose the contrast and definiton of the wood burning.

Ready to get started? We're going to begin with a "welcome" sign on a birch wood plank with routed edges. I'll take you from start to finish through this project; then I'll show you the high-lights from five other works, so that you can easily create them from the patterns in the book.

ABCDEFG

HIJKLMN

OPQRSTU

VWXYZ

abcdefghijk

lmnopqrstu

vwxyz

1234567889

Patterns

Pattern is reduced 84%.
Enlarge 122% for original size.

Welcome to Our Home

Pattern is reduced 84%.
Enlarge 122% for original size.

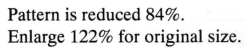

Home is Where the Heart is

Pattern is reduced 84%.
Enlarge 122% for original size.

Silent Huntin'

Gone Fishin'

Welcome To Our Home

Pattern is reduced 84%.
Enlarge 122% for original size.

Welcome Sign

Your first step in wood burning is to prepare the wood. Here I'm sanding with the grain on my board, using a small sanding disk of fine-grit paper.

In order to transfer my pattern to my board, I first trace the pattern onto a piece of tracing paper, using a number-two pencil. For your first wood burnings, trace as precisely as possible so that you don't have to "wing it" while using the burner, filling in blank spots without a reference point. You'll also find that tracing gives you an opportunity to become familiar with your picture, so that when you burn you'll have a better idea of where your lines are going—and, therefore, better control.

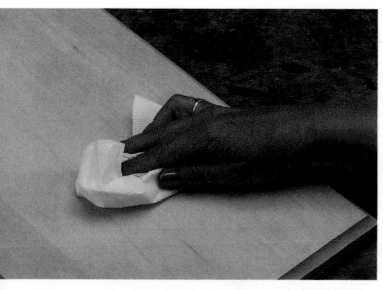

When the board is perfectly smooth to the touch, I wipe off the dust with a paper towel, so that it doesn't interfere with my burning.

Your next step is to transfer the pattern onto your board. Place the tracing paper on the board and trim it to match the wood size, so that you don't have to deal with a messy overhang.

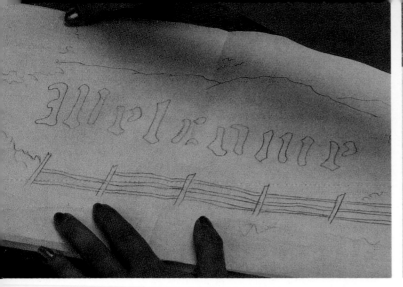

Tape the tracing paper to the board, either at the top or the bottom. Doing so holds the tracing in place but allows you to lift one edge in order to check what you've transferred. (And you will have to check—it can be surprisingly easy to lose your place while you're transferring the pattern.)

As you trace, press firmly, but not as if you want to plow the wood. Pressing too hard will put a dent in the board's surface.

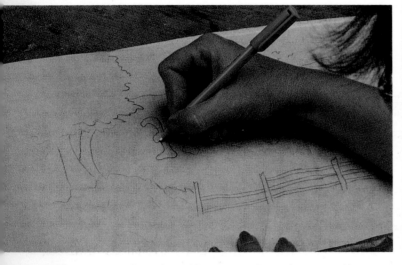

Next , place a piece of graphite paper underneath your tracing paper, against the board. Trace the pattern again, transferring it to the wood. To help you keep track of what you've traced, use a colored pencil or a firm felt-tip marker rather than the pencil.

Although I'm using a purple marker, I still missed part of my drawing. I see that right away when I lift the tracing and graphite papers. I'll slip in the graphite paper and trace the shape that's missing.

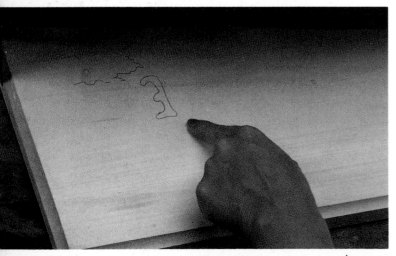

It's possible to waste a lot of effort by tracing with the graphite paper facing the wrong way! Prevent this problem by tracing just a little bit at first, lifting your tracing paper, and making sure that you've got a line.

With the tracing complete, I'm ready to start burning. I'm using a solid-state unit at a low temperature.

Begin with the letters that spell "welcome". They're solid and dark; you don't have to worry about subtle shading and mistakes are easy to cover up. The first step is to outline the letters. For this fine line work I'm using the writing tip.

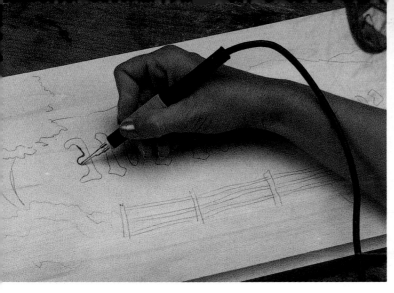

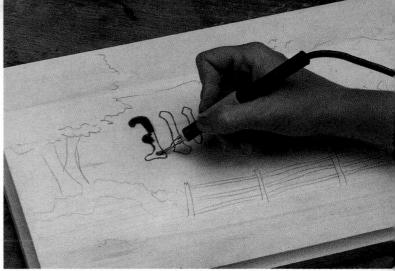

The burning may go faster than you think. Follow the pattern, keeping the burner moving so that you don't get a "burnover."

Here's an example of what can happen if you crank up the heat on your solid-state unit too high. The wood catches fire and you get a burnover. At the very least, heavy shading will put a depression in the wood—acceptable only in all-black areas like the letters.

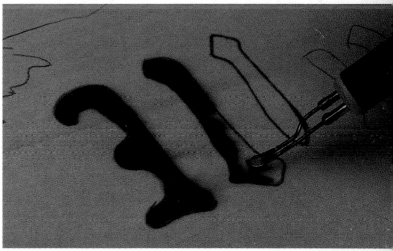

Work toward producing smooth, continuous strokes. If you start and stop, you'll get bumpy, uneven lines, as I've demonstrated here within the outline of the first letter.

Here's a close up of a burned letter. I don't like the hazy burnover at the edges . . .

Fill in the lettering with the shading tip (or use the broader face of your single-unit tool). Use a circular motion or burn using vertical strokes—whichever is most comfortable for you .

So I clean it up using the edge of my sanding disc.

The lettering is nearly finished. Strive for this look in your work: solid fill, with crisp edges.

Keep the left side—the highlighted side—light by moving the burning tool quickly.

With the lettering done, changing to the writing tip and move on to the rest of the board, starting in the foreground and working back. (If you work from back to front, you'll find yourself burning over your work, making for unnecessary dark areas and crossed lines.) Determine from where you want the light to be coming; in this example, it's coming from the left. That means that the shadows will fall to the right. Start with a dark area: the right side of the fence post.

I like to work on a fence section by section, so I've burned the second post just like the first one, and now I'll go back to fill in the rails.

Work in vertical strokes, following the "grain" of the fence post, to texture it. You'll notice that the grain of the wood provides variation in the lights and darks, adding character.

Because of the way the light falls on the rails, the top of each one will be lighter than the bottom.

I've darkened the section of the rail just to the right of the post, where the post casts a shadow. The rest of the rail is fairly even in tone.

Add some variation to the texture of the wood by slowing your burning to produce dark spots. They look like natural depressions in the wood.

I shade each successive rail in basically the same way: shadow behind the post, darker on the bottom.

Take a break during burning to step back and look at your work. Here I notice that I could use some more shading to bring up the contrast in the fence.

When I burn the next set of rails, I rotate the board to make it easier to produce a smooth, straight line.

I add the shading by switching to the shading tip and burning along the right sides of the fence posts.

With the writing tip in place again, I move on to the bushes. Bushes are fun: they're basically a collection of squiggly lines. Keep your strokes loose, especially in the areas where the bushes are in sunlight.

The bottom of the line of bushes is jagged, so that I can pull up grass into the line and give it a natural appearance.

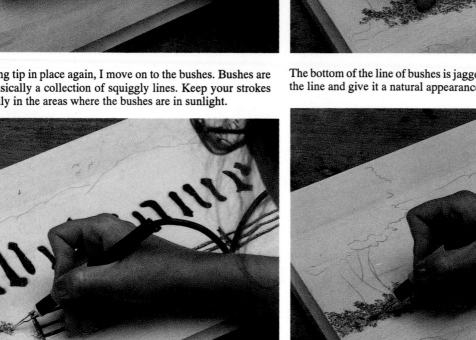

I make my lines tighter toward the right, because that's where the bushes are in shadow.

Now, with the entire bush laid in, I switch over to the spoon tip to darken the shadowy areas quickly and provide better contrast. (Contrast is the key to making your wood burnings appealing.)

I scribble the tip around the lighter face of the fence post, helping to set it off and to define the shadowy section of the bushes.

With the bushes to the left of the fence complete, I "squiggle" in the bushes on the right-hand side.

In this picture the contrast between the completed bushes (to the left) and the roughed-in bushes is clear.

With the shading tip in place, I add shading to the bushes. If a spot gets too dark . . .

The trees are in the background and should therefore have less detail than the fence and bushes. I'm burning them in with the shading tip to produce a softer, less-defined line.

. . . camouflage it by adding other areas of darkness and blending them together with lighter shading.

Use the same scribbly stroke for the leaves that you used for the bushes, burning more slowly in some areas to produce contrast and add fullness to the tree. (Think of what a tree looks like: some branches extend farther than others, so there are shadows within the foliage as well as highlights.)

The tree trunk is "behind" the tree foliage, so I burn it in next. I'm using the writing tip to help me texture the tree.

The mountains, done with the shading tip, are quite soft because they're way in the distance. However, they still need contrast–shadow on the right and lighter area on the left–to take shape.

As with the fence posts, the right-hand side of the tree trunk is darker than the left.

The light areas in the mountains are done with fast, loose strokes. I burn in the dark areas by moving the tip slowly, in vertical strokes.

I move on to the tree on the right, working all the way out to the edge of the board.

Even on the dark or shadow side of the mountain ridges, there are variations in the values. I add some darkness in these areas and the mountains really start to pop out.

Because the very small, far peaks are way in the background, they don't have much detail. They're basically dark on the right, with a loose, scribbly left side.

With the mountains complete, I switch to the writing tip and draw in some birds. Birds add life to a picture, yet they're very simple to burn. They're basically a stretched "m."

The last area that has to be burned is the grass. I use the shading tip because I want the lines of the grass loose and soft-looking.

To add texture to the grass, burn some areas darker than others.

Burn grass by making vertical strokes.

When you've burned grass in from edge to edge with the shading tip, go back with the writing tip (or the edge of your single-unit tool) and put in some defining lines. To prevent yourself from burning in a lot of dark marks, keep the burner moving up as you bring it into contact with the wood.

"Plant" the fence posts by burning grass up around the bases.

Here's the Welcome Sign complete. Now that the burning is complete, it's time to put on the finishing touches.

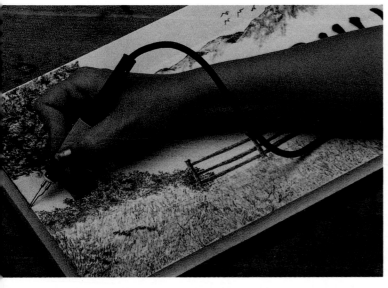

I don't intend to frame this board, but I am going to paint the routed edge. To keep the paint from seeping onto the wood, I take the writing tip and burn a line around the face of the artwork.

Acrylic paint will dry quickly, though, so I work steadily and don't let the paint get above this line on the brush. It can set up near the ferrule and ruin the brush in fifteen minutes.

I like the painted edge to match the burning, so I use burnt-umber acrylic paint. I use acrylic paint so that I can thin it and clean it up easily with water.

Start at a corner and work down the edge; you'll find that the grain-end of the board takes a lot of paint.

27

The finished edge sets off the board nicely. Now I'll let it dry at least 45 minutes before applying the varnish coat.

I varnish the board to prevent it from darkening (as wood tends to do in sunlight), because on dark wood the burning loses impact. Pour the varnish on . . .

I let the varnish dry for about 30 minutes, sand the board lightly, and apply a second coat.

. . . and smooth it from edge to edge with a paint brush.

Not only does the varnish protect your burning, it really brings up the contrast.

28

Welcome Plate

Our next project is a plate with a floral design; this is what it will look like when completed.

For this job, I'll be using a slant board (a smooth piece of fiberboard with a one-inch wood strip nailed on the back edge) to help lift the work, make it more visible to me, and to take some pressure off my wrist.

I've already traced the pattern onto the plate and burned in the lettering, "Welcome to Our Home," using the same techniques I described for the first board. Now I'm burning the floral pattern, using the writing tip. I start by burning the flowing line that forms the main "stem" of the flowers.

Starting on top of the stem line, I burn on the individual stalks for the flowers. I've paired them, turning the stalks in each pair toward each other, and turning the pairs away from each other.

Fill in a flower for each stalk.

Burn and fill in five loops for each flower, placing a flower at the end of each stalk.

To achieve a lacy look, stagger your placement of the stalks on the underside of the stem with the paired stalks above.

Place a five-loop flower on the end of each stalk.

When every flower stalk has leaves, go back and burn in shorter lines between the stalks. These will have leaves only—no flowers.

Add leaves to each flower stalk. I like to vary their placement on the stalk, making some opposite each other and some catty-corner in order to keep the pattern open and interesting.

I burn the leafy stalks onto the top side of the central stem first; then I add the tendrils to the bottom.

Add a leaf or two at the tip of each flower to complete the lacy look.

When you're satisfied with the way your plate looks, it's ready for varnish.

Home is Where the Heart is

The knot hole and the bark on the edges of this plaque give it a great rustic look. I'm going to use it for a mountain scene, with a stand of pines to the right of the slogan and a line of mountains behind it.

I've sketched out my pattern and have filled in the lettering. Now, using the writing tip, I'm burning the pine-tree trunks. I start by putting a dot at the top of the tree, then moving my tip farther down and drawing up to the dot. This method helps me to narrow the line as I go, just as a tree trunk narrows as it goes up.

Then I go back and start filling it in, working from the apex of the triangle that the zig-zag forms with the trunk back to the trunk line.

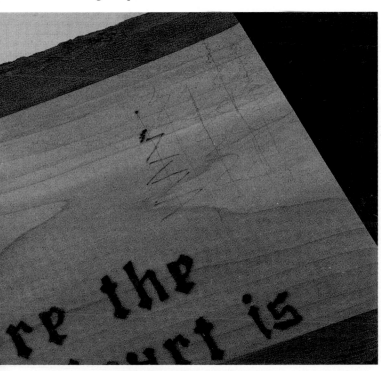

The zig-zag line of the pattern helps to produce a natural-looking pine tree, guiding you to stagger the branches and keep them from getting too large. I burn the zig-zag in first.

"Squiggle" in the pine boughs, moving the burning tip in a tight, random pattern.

I use short strokes to fill in the narrow top of the tree. The grain of the wood helps to produce nice lights and darks.

Step back from your work and take a look, then darken the shadow areas. My light is coming from the right, so I want the darker areas to the left. Using the shading tip helps me to do the work quickly.

Burn each tree the same way, working to the edge of the board.

I've added some soft, smooth strokes beneath the trees to pull them into the ground.

Burn in the mountains just like you did for the first project, except here make the right-hand sides of the peaks lighter than the left.

Loose, soft strokes give the mountains an impressionistic look.

The darkness of pine trees balances the darkness of the knot in the wood; the whole picture pulls together.

Went Huntin'

I've sketched out my next work: an elk, aspen trees, and a "Went Huntin'" stencil. The challenge here is to give the elk a fur-like coat.

Start with the antlers; the light source is from the right, so the left sides will be darker. This intricate work requires a low heat setting.

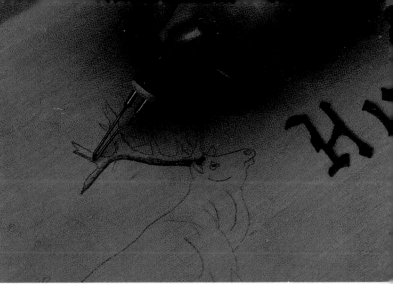

The left, or back, edge of this prong of the antler is dark, too. The key to making your elk look real is to keep your lighting source consistent.

Burn a line up the back of the antler with the writing tip, using a smooth stroke.

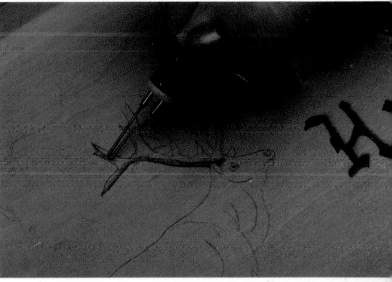

A line burned darker on the prong gives it a realistic shape.

Fill in the antler, keeping the back edge dark.

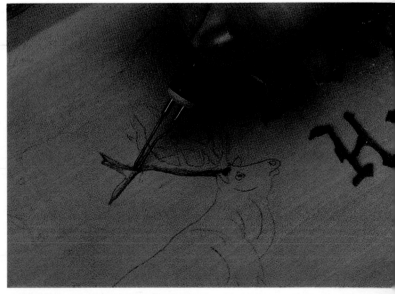

Shade each section from dark to light: here I'm shading the little corner formed by the dark line within the prong.

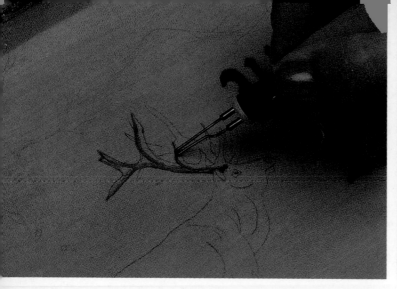

I move down the antler, burning in the dark areas first.

A solid dark line on the elk's left antler helps to separate it from the one I just burned.

With the back edges dark, I'm ready to shade in the front of the antler.

I keep the shading simple on this antler, so that it seems to recede behind the first one.

The grain of the wood helps give a natural, variegated appearance to the antlers; the smooth, light burning strokes add the values.

Put in the eye with a single dot.

The nostril is one dot; the mouth is a simple line.

. . . and the nostril.

When you burn in the face, put in the highlights first. "Scribble" lightly with your writing tip around the jaw.

The darker areas of the face and the body of the elk can be done with a shading tip. Here I'm defining the line behind the jaw.

. . . the tip of the nose. . .

Burn in the hair on the face, imagining what an elk's coat is like: its texture, and the direction in which the hair grows. (If you don't know what the coat is like, look at some photographs of elk.)

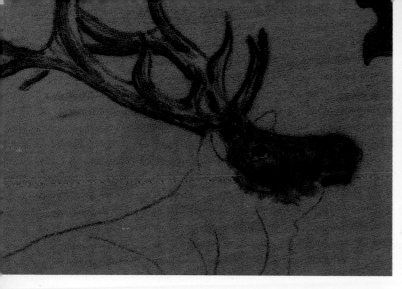

Here's a shot of the completed face, showing the direction of the burning strokes.

Behind the highlight is another section of dark hair. The strokes on the neck are all basically in the same direction: down, on a slight diagonal.

I find it easiest to work on the elk from front to back. I've filled in the ears, leaving a highlight inside, and now I'm burning in the long hair on the underside of the neck.

At this point on the neck—about one quarter of the distance from the top—the hair changes direction, and the burning strokes are up and to the back.

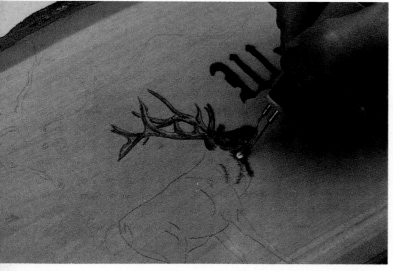

The hair along the underside of the neck is shaggy and dark, with lighter hair behind it (I've marked the lighter area with a rough line).

An elk has a slight depression behind the top of his shoulder: I'm marking it with a darker burn.

Set off the back of the shoulder with a shaded area, too.

Behind the highlight, the leg gets darker. I've made a mark to show where the shadow should appear.

Bring your darker strokes around under the bottom of the shoulder, to help give it its full shape.

Burn the darkness down the back of the leg.

Define the front of the leg with a light stroke.

By leaving the front of the leg light and making the back dark, the leg takes on shape. I leave the foot area undefined, because I want to be able to burn grass up around the elk's legs.

41

Tie the leg and shoulder into the neck with a few dark strokes of shaggy neck hair.

Fill in the haunches by first making a guiding line along the back of the leg.

Use the same type of stroke to burn in the hair on the elk's belly. Because the body of the animal is round, let your burning become lighter as you work your way up his sides.

Just as with the front leg, the front surface of the hind leg should be lighter than the back. Make quick, smooth strokes, blending in toward the shadow side.

The back has just enough color to define it.

The legs of the elk on the other side of his body should be darker than the shaded areas of the legs in the foreground, so that they recede.

Leaving an unburned line between the body and the far leg helps to keep them separate.

Extend the shading from the dark areas, using quick strokes to lighten the burn and produce contrast. Burning in hair is like burning in grass: use up-and-down, scribbly strokes.

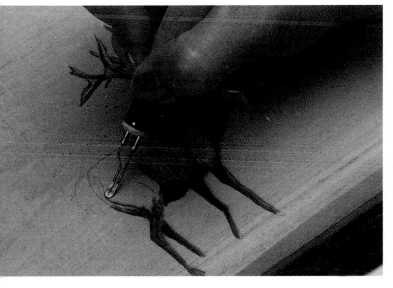

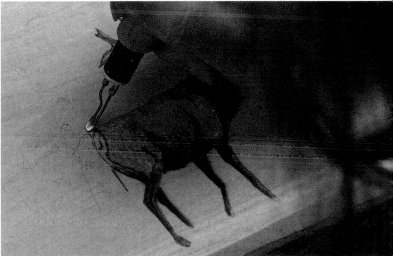

Burn in the lines of the haunches. There are two: the line that defines the edge of the haunch, and a line toward the center of the body that defines the animal's hindquarter muscle.

A dark tree behind the elk's light tail will help it to show up. These trees are aspens; they have vertical lines on the lower trunk, and as you go up the tree, the knot holes are on the horizontal.

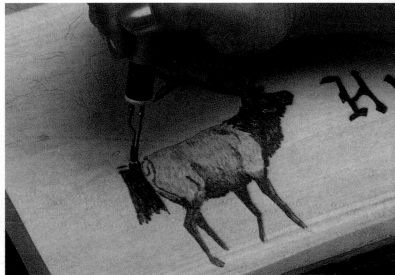

Round out the haunches by burning in the pattern to define the tail.

The elk starts to stand out as the tree takes shape.

The knot holes are a collection of dark horizontal strokes.

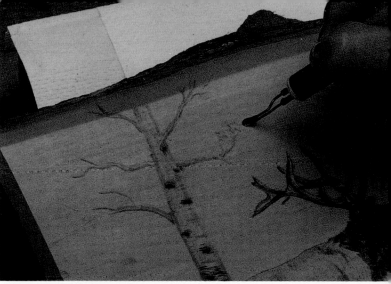

The leaves are done in the same manner as those of the trees on the first board: with loose, scribbly strokes.

Fill in the rest of the trunk with light vertical strokes. Leave the base of the tree ragged, as you did with the legs of the elk, because you'll be working grass up into it.

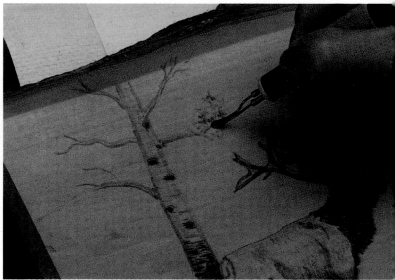

Add some dark areas to bring contrast—and fullness—to the foliage.

Add the branches, keeping the dark, shaded areas to the left, as you did with the parts of the elk.

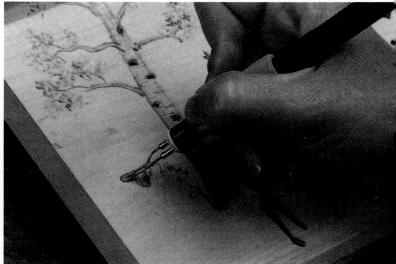

Fill in the bushes behind the aspen, making them darker than the tree leaves, which are farther away from the front of the picture. Keep the line of bushes away from the trunk, so that the two objects don't blend together.

44

Move on to the mountains, burning the dark sides of the peaks on the left.

The pine trees are done with tight, dark strokes. Work off a zig zag if you don't feel comfortable free-handing, as I'm doing here.

On this board, the grass won't be as dense as it was on the Welcome sign: a few clumps here and there suggest grass; grass under the trees and the elk help to ground them.

The completed board. It should be varnished several times, with a light sanding between each coat.

Gone Fishin'

This next board, a "Gone Fishin'" sign, will be done in silhouette. That means burning a fairly large area. The trick is to keep the value even and the outlines crisp.

I wanted the edge on this board painted. Painting the straight (non-routed) edge later without getting paint on the burning would be difficult, so I've put the paint down first. Then, using the writing and shading tips, I've filled in the lettering. Now I've started to fill in the silhouette.

Working with a high-heat setting (not a problem with the single-unit tool) causes a burnover. You can see how hazy the edge of the tree is.

I correct the burning by sanding back the fuzzy edge.

The direction of the lines adds interest to the burning, just as brush strokes add interest to a painting. When I burn the tree, I do so with vertical strokes, suggesting the direction of bark.

Switching to the writing tip, I burn in the details: the fishing pole and grass along the edge of the lake. The rule when you're working on fine details such as these is to go in light and burn over the line a couple of times. This helps you to control the level and direction of the burn.

Notice the cross-hatch pattern of the grassy bank. Even strokes prevent the burner from gouging a hole into the wood.

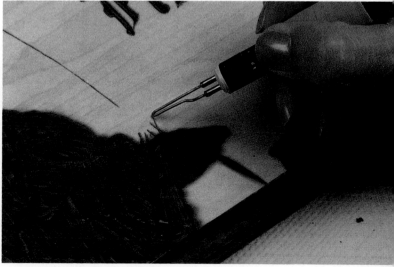

I draw in the grass by pulling the burner up from the bank.

Here's what the burning looks like with the silhouette complete. Now all the picture needs is the water.

Because I don't want the water to blend into the bank, I'm burning it in lighter than the surrounding area. I use horizontal strokes to imitate the natural appearance of a lake, leaving some light areas to suggest sunlight falling on the water.

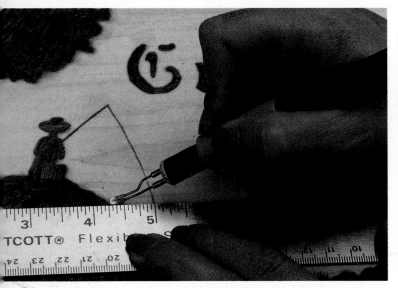

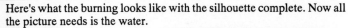

There's no fudging with the horizon line on a body of water. It has to be level. To make sure mine is, I burn in the line using the shading tip against a metal-edged ruler. (If you use a plastic or wood ruler, you're going to regret it!)

The edges—where the lake touches the banks—are darker than the body. Where the fishing line drops into the water, I burn in a few dark, curved lines.

Add a few birds to the sky and the burning is ready for varnishing.

Picket Gate

This welcome plaque with its simple gateway and stone path will introduce you to two new textures, wood and rock.

I've transferred my pattern to the plate and have burned in the lettering.

Using the writing tip, I burn in the shingles on the roof. They're basically overlapping rectangles and any unevenness adds character.

I finish up the shingled side of the roof by darkening the line along the back.

Filling in the shaded underside of the roof is a matter of drawing roughly horizontal, parallel lines. Again, don't worry if your lines aren't perfectly straight; the unevenness brings a folksiness to the work.

Outline each picket , tracing them with the writing tip as you would lettering.

The hinges are two dark dots. (Little details such as these really bring charm to a picture.)

The dark lines between the pickets of the swung-back gate suggest shadows falling on the horizontal boards between the pickets.

I switch to the shading tip to darken the area under the roof.

Contrast is important to any wood burning, so I use the shading tip to darken the left sides of the roof uprights.

I also shade the left sides of the pickets.

Using the writing tip, I "scribble" in the grass that will surround the stepping stones. When I burn in the tops of the rocks, the grass will seem to set them right into the ground.

I like to do the rocks with the shading tip, making a broader, softer line that contrasts nicely with the upright grass. One dark line shades the undersides of the rocks.

And another, narrower line defines the top.

53

The plate is complete.